On the Origin of Humankind

On the Origin of Humankind

ILYA GLADKIH

Archway Publishing books may be ordered through booksellers or by contacting:

Archway Publishing
1663 Liberty Drive
Bloomington, IN 47403
www.archwaypublishing.com
844-669-3957

ISBN: 978-1-6657-7096-5 (sc)
ISBN: 978-1-6657-7097-2 (e)

Library of Congress Control Number: 2024927188

Print information available on the last page.

Archway Publishing rev. date: 12/27/2024

Preface

Our understanding of humankind's origin has remained unchanged for over 165 years—until today. It's difficult to understate the significance of the findings in this book. I believe this book will change the reader's perception of humankind's origin. I'll do my best to provide compelling evidence for my findings by the end of this book.

I want readers to understand that this book began as half a joke. I wanted to write it in my free time, having fun with only a limited time of two weeks. However, I invested months in research and book polishing during the project. Because I wanted to be as clear as possible. Despite my efforts, even after six months of work, I still find some sentences with inaccuracies, grammar mistakes, and unclear meaning, which makes me mad. I had to learn to enjoy the daily effort. I also had to learn how to break work into small pieces, which I can manage.

Even though this book is incomplete and not as polished as I wish, I still want to publish it. This book is significant because it can impact humankind's understanding of the past and hint at some future discoveries. In the future, I want to focus on writing other books that benefit people emotionally and materialistically. Please don't anticipate accuracy, references, or deep fact-checking because I am not a scientist. In this book, I'll try to answer the question of the origin of humankind, and spoiler alert: it was not an accident. Even though this book is not as polished as I wish, I believe readers would find the method described in this book fascinating. English is my foreign language, and I mostly rely on Grammarly to correct mistakes. If readers would see grammar mistakes in my book, please forgive me, and feel free to send angry letters to Grammarly developers. I am sure they will find constructive critique very informative. Of course, I am fully responsible for the content and errors related to the ideas and their representation.

Before I begin, let's understand what information has been collected by scientists that can help in the search for humankind's origin. There is so much information that it is unbelievable: for example, knowledge about DNA, chemistry, and human biology. For those who are unfamiliar, DNA, in simple words, is a molecule that contains the genetic code of a living organism. The word DNA is often used to describe a living organism's complete or partial genetic code. Until this point, the most popular theory was the theory of evolution by Charles Darwin. The book "On the Origin of Species " by Charles Darwin was published in 1859. This means the theory of evolution has existed for over 165 years! The more I learn about Charles Darwin, the more I'm fascinated by him and his intelligence. To truly understand how innovative his book was, I want to emphasize that DNA had not yet been discovered when his book was written. People had no idea how species were formed. This is what makes Charles Darwin such a fascinating figure. He collected massive data on the species from different geological locations. He also collected data on discovered fossils and understood how they changed throughout time. The dating fossil technique was a groundbreaking discovery at that time. This astonishing and humongous work resulted in speculation that species developed through variation and natural selection. Charles Darwin saw that domesticated animals such as dogs had a significant variation, which resulted in hundreds of breeds. Then, he took a step forward and speculated that the same thing happened in the animal and plant kingdoms. He assumed that variation created different species through natural selection. It was remarkable and enormous work. Later on, scientists discovered DNA. More importantly, scientists have found gene variations through mutations. Scientists were also able to find out that the DNA of a human was very similar to the DNA of a chimpanzee. What was also important was that scientists were able to discover traces of retroviruses in the same places in the DNA of humans and chimpanzees, which indicated that humans and chimpanzees had the same ancestors.

On the other hand, some parts of the theory of evolution are difficult to explain. The complexity of the solutions in the human body has an extremely low chance of occurring by mutations. For example, an eye is a complex structure that cannot be broken into individual components that could be selected by natural selection. The same goes for the first cell. Scientists, to this day, can't confirm how the first cell appeared. This is because multiple very complex structures must appear simultaneously. One of the essential components of any life form, like bacteria, microorganisms, plants, animals, and humans, is the ribosome. A ribosome is a micromachine that converts RNA code into proteins. RNA, in simple words, is a molecule containing genetic code copied from DNA and is used in various ways, one of which is to transfer genetic code to ribosomes to produce proteins. A ribosome from the bacteria E. Coli is built from 22 different proteins. It is unlikely that those 22 proteins would appear simultaneously with RNA code describing them. That is an unbelievable accident to have proteins and RNA code for them at the same place and time. Another implausible factor is that even if suddenly, by accident, 22 different RNAs that describe ribosomes would appear, those RNAs are still vulnerable. Those RNAs will break after some time, and there must be micromachines to copy and store them. The number of weak points in the theory of evolution is staggering. This is why some people do not believe in the theory of evolution, even though it has been beneficial since 1859. The theory of evolution was useful for at least 165 years. The theory of evolution is useful even today, for example, in studies related to antibiotic resistance. In one study, it was discovered that bacteria could acquire incredible antibiotic resistance in eleven days. The amount of antibiotics that bacteria could resist by the end of the study was a thousand times more than the amount of antibiotics that could kill the bacteria in the beginning.

Some people advocate for a theory of intelligent design. I watched YouTube. There are arguments in favor of this theory. However, I would say that arguments favoring intelligent design theory require more intelligence. They were based on speculations or facts that couldn't be fully proven. One guy even brought documents from the court, which he lost and argued against arguments in this document. In other cases, some people advocating for intelligent design tried to vilify science and disregard scientific methods. In general, the argumentation methodology that was used resembled pseudoscience. In the end, one chemist was asked if he had a method that could distinguish between intelligent design and random accidents, to which he answered that he didn't have such a method. The gaps in intelligent design theory were more significant than in the theory of evolution. More importantly, intelligent design theory contained no information that allowed scientists to foresee future events and discoveries like the theory of evolution. Even with those gaps, I respect those who advocate for intelligent design because they truly believe in what they do. Eventually, this book may add a little twist.

This book investigates whether humankind was created by accident. I will present a compelling method for testing this theory. As I've already hinted, the answer is not what you might expect.

Here, I will write some information about myself. I am not a biologist or scientist. I worked as a software engineer in one of the leading high-tech companies in the world. In total, I worked in high-tech for over a decade. Even though I don't have a deep knowledge of biology, I must write this book. As readers will later understand, this is because I'm an engineer—to be more precise, I began as an associate electrical engineer. This will play a crucial role in this book.

My First Doubts in Evolution

In the age of technology, so much information is available that people find it difficult to navigate. Unfortunately, in most cases, this information is inaccurate, misrepresented, or serves someone's agenda. The same goes for many available books. Some cults try to convince people that they are only true faith. After making several mistakes, I understood I needed tools to verify the information's truthfulness. For this reason, I developed mini-processes to confirm the information. This process includes several steps. At first, I check the sentences to see if they follow a formal logic. Then, I would make a fact check and verify that the sources are worthy of trust. Then, I would simulate it in my mind and imagine how it's supposed to work. The next step is the most important: I would bring theory into the world and see if it works. Of course, I cannot include every possible check for all the information I hear or read. However, about 90% of incoming information fails formal logic and fact checks. There is often no need to verify information because some cues indicate it's a lie. For example, when someone tries to modify the meaning behind words or turn off the natural defense mechanism of logical thinking. The first thing people who want to trick others say is that, somehow, feelings are superior to logic. Those things likely indicate the use of manipulation tactics. It seems that the theory of Charles Darwin is pretty accurate on logical and factual levels. However, I haven't done a deep fact-check. I spotted something interesting on the third level when I tried to simulate how this theory would work in practice. In my mind, I was attempting to model mutations that would occur during the natural selection process.

I have found it surprising that when a mutation occurs, there are three options. The first option is when the mutation is successful. This successful mutation does benefit the individual. With time, this mutation will become more popular and, after a while, will replace the existing mutations. In the second option, if the mutation is harmful, this individual will likely die, and this mutation will be eradicated. However, I discovered that there is a third option, which occurs when the mutation is neither beneficial nor harmful. In that case, the proportion of individuals with this mutation will stay the same. For example, if a village has precisely 100 people and one of them has neither beneficial nor harmful mutation. With time, after the village's population grows to 1000 people, the percentage of people with this mutation will stay the same, which means ten people. This is not what is seen in real life. Humans have, on average, 32 teeth in their mouth that are precisely located, so there should be many mutations when teeth are located somewhere in the body. In some cases, this would be deadly. In other cases, this mutation would be neither beneficial nor harmful. In the worst case, teeth would grow somewhere with some survival benefit not intended by design. This means people should have random teeth growing somewhere in the body. This is not what is seen nowadays.

Darwin's book addressed the absence of transitional mutations. However, the assumption in the book contained a mistake. Although the book of Charles Darwin describes variations in species, it is now known that DNA mutations are behind those variations. This is why I use the word mutation instead of the original description. The assumption in the book was that only beneficial mutations would be preserved because they would replace all other mutations. The fact is that there is a huge amount of genes in DNA. Many of those genes could be removed or altered without life-threatening consequences. This fact makes the explanation for the lack of transitional mutations unreasonable. When one mutation is beneficial and spreads, other mutations in other parts of DNA can also spread independently. This means there should be some proportion of transitional mutations in the population, even if they are neither beneficial nor harmful. This is a significant flaw in the cornerstone of the entire theory of evolution. The whole theory breaks apart without a working variation and natural selection process. This is when I began to suspect the theory of evolution is wrong. However, just thinking about it was not enough for me. I wanted to find a method that would prove with nearly 100% accuracy which theory is correct. Without wondering, I just wanted to know for sure, and I found the answer.

Let's check evolution from a mathematical perspective: one condition must happen to make one gene with an advantage in survival to replace another throughout the entire population. The cause of death, which mutation solves, must eradicate the entire population of species that don't have better genes. Otherwise, a better gene would not expand throughout the entire population. To understand this, let's say there are two groups: the first doesn't have a nail on pinky, and the second does. The group with a nail would expand throughout the entire population only if the group without the nail is in reproductive decline. This means that for each one of the parents, there is less than one child. Those conditions can happen only if there is a change in survival requirements or if a group with a nail on a pinky causes the reproduction decline of the group without a nail on a pinky. The only other condition in which nails can appear in the second group is when this gene has more than fifty percent to pass to offspring. This condition can only be caused by computation results. This example can be made on teeth, ligaments, nails, fingerprints, etc. The results are the same for many body parts, which have an insignificant effect on survival.

I put a lot of effort into making this book beneficial for readers. If you find it fascinating, please tweet, comment, TikTok, or share it with friends, family, and colleagues. Any help with making this book reach greater audiences is very appreciated. Because this will allow me to produce other important and impactful books. I am pleased to say "Thank you!" to anyone who helps me. I deeply appreciate your help. I can't state how important it is for me. And I will do everything I can to make this book worth reading.

Asking the right question

After I understood that evolution could not explain the perfect human design that is seen nowadays, I started to investigate the origin of humankind. I wanted to find a way to prove humankind's origin with 100% accuracy or very close to that, regardless of what kind of origin it could be.

At first, I started to look at the initial complexity of the first cell. There was some gut feeling about it. Eventually, this was a false path. However, I found important information that helped me in my search for the origin of humankind later on. The common evolutionary explanation was that a first cell was created in some soup of molecules. The fundamental problem with this explanation is that molecules break faster than assemble into complex structures. However, micromachines always assemble molecules into complex structures inside the cell. The question was if basic physical laws could produce micromachines like ribosomes and RNA code without intelligent intervention.

The way I looked at the problem was that there were limited future possibilities after the Big Bang. There was an infinity of possibilities. However, this infinity does not include all possibilities. The best example could be in numbers. For example, there are no rational numbers in an infinity of whole numbers.

Another example of how this could be imagined, let's say there is a long glass. On the bottom of which is sand, and in the middle is a separator made from wood. On the top, there are small stones. The glass is sealed from the top. No matter how the glass is shaken, the sand will not mix with the stones because of the separator. When glass is shaken, an infinity of possibilities exists for each grain of sand to be placed in space. The same infinity of possibilities exists for the stones. The stones and the sand would never mix without external intervention. I imagined that the same goes for molecules. They would not form complex structures without external intervention.

However, I faced an issue. I could not prove that the molecules would not form those complex structures without external force driven by intelligence.

From a mathematical perspective, intelligent intervention is indistinguishable from random force. This fact made it almost impossible to rule out the involvement of the accident. Although the odds of such an accident were minimal, this challenge was significant enough to cause me to look in another direction.

Because of my engineering background, I have noted that species are very similar to engineered products. For example, the evolution of cars throughout the years is suspiciously similar to the evolution of species. As an engineer, I saw a similarity between complex solutions in the human body and engineering solutions. As I mentioned, my engineering background is crucial in this book. Solutions in the human body and engineered products are considered good from an engineering perspective. It was unreasonable to think that natural selection would produce solutions according to engineering considerations. For example, the arm is engineered with precise fingerprint limits. There is a long list of solutions in the human body that I noticed were made similar to engineered solutions, like ligaments and muscles in the arm, which are located exactly where they are supposed to be from an engineering perspective. Muscles are located in the forearm and not near the fingers, which allows such an elegant and efficient hand design.

The way that all ligaments are connected to fingers without exception or duplication just doesn't make sense from the selection process point of view. The way the human body is made is fantastic, combining symmetrical and non-symmetrical body parts, just like a car is engineered. Those things were similar to engineering products and not to selection process results.

This kind of resemblance caused me to ask myself: "Can I calculate the odds of a human being engineered?" I was surprised that the answer was "yes", and it was straightforward to do. At first, I broke it down into three different questions.

1. Can I calculate the odds of a car being engineered compared to a rock?

I want the reader to understand that physical laws form the rock, which would lead to the answer to the first question. Physical laws have limitations on the variations they can produce without intelligent intervention. I found this principle while investigating the origin of the first cell. It's amazing that now I am using this principle in such a surprising way. This principle says that physical laws without intelligent intervention can only produce a limited number of outcomes. Engineering, however, can consistently deliver results that match complex specifications. This means it's possible to measure the difference between something engineered and something made by the laws of physics without intelligent intervention.

2. Is there a measurable difference between the engineering and selection processes?

Engineering solutions can differ, but the approach is always the same. Different trade-offs could exist within the specific design. However, the same engineering process would always be behind any design.

The selection process can only recognize solutions that provide a measurable difference in the solution's survivability. The engineering solution would be completely different because it would depend on the trade-offs and processes defined by the engineer. This means there must be a difference between engineering and selection processes.

3. What are the odds of human being engineered?

There is a significant difference between "What are the odds of human being engineered?" and, for example, "What are the odds of human being produced by intelligent design?". I think the choice of the word intelligence in the name of theory has exponentially increased the difficulty in proving it. This is because it is hard to define what intelligence is. Is this intelligence when deciding to eat a burger before or after noon? In general, it is, but is this difference measurable? The same goes for design. Design is when you pick one color over another. The problem is that no one can reliably measure the difference between designed things and those which are not. However, when something as specific as engineering is discussed, it's possible to calculate the probability of engineering effort with insane accuracy. Because it's possible to measure the difference between the measurements of engineered products and those that are not. The question: "What are the odds of human being engineered?" is extremely specific. Asking the right question, in this case, gives 95% of the answer.

Method of Measurement

How exactly will I measure the probability of the engineering origin of humankind? Let's consider a simple example based on chess. Calculating the probability of playing against an opponent who knows the rules is possible by counting how many moves comply with chess rules. There are only two options: moving chess figures according to rules or not. If the game had ten moves, the result would be ten "yes" or "no" answers. Each move that follows the rules will reduce the chances of playing against someone who doesn't know the rules by 50%. If there have been ten moves, playing against someone who doesn't know the rules will be one in 1024. To make calculations easier, let's round this number to 1000. After playing two games while each has ten moves, then the combined odds would be 1 in 1000000. This number would increase exponentially depending on how many games have been played. This is one of many measurement options that could be applied. However, the number of moves can change for each game. The number of moves can be below ten, and it can be above 10. Another method could be applied to make calculations more consistent. It's possible to measure the characteristics of chess games that distinguish between opponents who know the rules and those who do not. For example: "Did the player finish the game?" or "Did he play according to the rules?" "Did he intend to win?" "Was he playing smart?".

After each game, the answer to those questions can be "yes" or "no". If there have been ten questions and all received a "yes" answer, the odds would be similar to the previous method. The odds would be one in 1024. If rounded, it would be 1000 for each game. This means that for each game, the odds would increase by 1000. If someone plays ten games with the "yes" answer for all questions, the result would be one in 10 to the power of 30 because each game increases odds by 1000, which is 10 to a power of 3. This means that the odds in favor of playing against someone who doesn't know how to play the game would be almost zero. In the same way, it's possible to measure the probability of something being engineered. All that is required is the ability to measure characteristics that distinguish between something that was engineered and something that was not. Engineering is a very specific and rigorous process. People can easily measure characteristics that would indicate if this something was engineered. What is more crucial is that almost every component in the engineered product would have the same characteristics that would suggest that it has been engineered. The odds would exponentially increase when those characteristics are measured for several components. What is even more important is that this method is reproducible. This means anyone can anticipate the same results for the components for which the odds have not yet been measured. Let's take, for example, a car and measure the odds for the engine. The indication would be that the engine was engineered. The same results are anticipated for the suspension system. It's obvious to expect the same results for other components in the car even though the odds of engineering effort for them have never been measured before. Measuring the probability of engineering effort for every car element is possible. This reproducible method of measuring the odds of components being engineered is the final answer to whether a human was engineered. This is because it cannot be faked. Everyone can test it for every part that belongs to humans, animals, plants, or things.

There is a question of whether measurements would differ for something that was engineered and something that evolved by natural selection. Surprisingly, the answer is yes. When someone is engineering a product, a long list of engineering considerations guides the design. One of them is safety, which is the paramount consideration. I'll give a short example: imagine if a car could reproduce itself like an animal and, with time, would evolve by natural selection. After many iterations, the car would significantly differ from the cars on the streets today. The car's survivability would be optimized. However, the safety of the passengers would be irrelevant. Any consideration that does not include the survivability of the car would not be optimized or even would be reduced. In front of the car, there is an engineered crumple zone. This crumple zone was not engineered in the first cars. After many accidents, engineers understood that the car's safety is less important than the passenger's safety. This caused engineers to iterate the car's design and make it safer each time. The front of the car is crushed to save the passenger from deadly deceleration. The car absorbs all the impact and decelerates as slowly as possible by absorbing the impact with the front. Engineers prioritize the safety of passengers, and they sacrifice the car. The same goes for other safety features that benefit passengers, not the car's safety. This means that anticipation is that the human body would have plenty of engineering choices that are not optimized by the selection process but by engineering considerations. A measurable difference is expected between things that were engineered and the things that evolved through the natural selection process. Following this logic, in the human body, measurements would indicate the origin of humankind.

Car Versus Rock

I want to calculate the probability of humankind being engineered. Let's check if this method of calculation works by answering the first question out of three: "Can I calculate the odds of a car being engineered compared to a rock?". This question relates to random rocks in the wild and some general cars used for everyday commutes. This question may seem too simple. However, it is important to understand the concept through the simplest examples.

Can I calculate the odds of a car being engineered compared to a rock? Does it have a clear purpose?

Regarding the clear purpose, let's see what the difference is. Is there a clear purpose for the car? The answer is obviously "yes". The car has a clear purpose, which is transportation. Let's see the answer for a rock. For a rock, the answer is "no". So, it matches the expectations of clear, measurable differences.

Does it have safety over-engineering?

Regarding safety over-engineering, the car includes multiple systems to ensure the safety of the passengers, including seat belts, crumple zones, and airbags. Despite such a variety of safety features, new features are developed in every generation. The safety of the car goes even far beyond that. The car's safety features are as big as the earth. All of the roads were developed because of safety issues. Once cars became common, transportation accidents began to occur. This is why humans developed roads, road signs, and driving rules. The entire road is a giant collection of safety features. Almost every part of the road is designed for people's safety. For example, highways include unique safety features like separation from incoming traffic and barriers at the road's edges, which do not allow the car to hit a tree or fall from the slope. This means that the safety of the car is over-engineered. Rock doesn't have any safety features.

Does it have precise limits?

Regarding precise limits, cars have precise limits because they are mass-produced. Mass-produced cars of the same model are almost identical, with some exceptions, like color or different tires. For example, windows, wheels, and blinkers are the same size and shape in all those cars. Rocks come in all shapes and forms and don't have precise limits.

Doesn't it have anything extra?

Regarding anything extra, let's take a look at the car. It has precisely two side mirrors, four wheels, and one driver's steering wheel. Engineers always consider the efficiency of material use, meaning a car must have a specific number of components. Components should be without redundancy if not required for safety. Lack of redundant components will reduce weight and make a car cheaper. Finding a car with five wheels on the road would be impossible when it needs only four. It's also impossible to find a car with three side mirrors when it was designed to have only two. Because rock doesn't have a clear purpose, it's impossible to define if it has anything extra.

Doesn't it miss anything?

Regarding missing components, an engineered car will not miss any wheels, side mirrors, or doors. When something is engineered, it must be complete. This means an engineer would ensure that no car window, seat, or headrest is missing. Because rock doesn't have a clear purpose, it's impossible to define if it misses anything.

Does its design involve multiple systems?

Regarding the involvement of multiple systems, any complex engineering solution is built from multiple systems that work together to achieve a common goal. A car is a combination of multiple systems integrated into one solution. A car has several systems that work for a common goal: fuel, electricity, oil, and cooling. Those systems have very complex interconnections that work flawlessly and coherently. In the rock, there are no multiple interconnected systems.

Does its design include emergency options?

Regarding emergency options, the car's design includes multiple features that are only applicable in emergencies. Those features were already mentioned, like the crumple zone, seat belts, and airbags. The car also has various other safety features, such as cruise control, autopilot, or an automatic call to emergency services in case of a crash. Rock, of course, does not have any of those emergency features.

Does its design work in all edge cases?

Regarding edge cases, the car can drive under very different conditions. The car also allows the driver to manage all the different situations that can occur on the road. So what happens if the car is low on oil, missing some water, has a malfunctioning sensor, or is low on gas? In those cases, the panel in the car indicates those alerts to the driver, allowing him to respond properly. Nowadays, cars also have complex sensors that provide additional information about these issues to the technician. With the proper equipment, technicians can see the status of the car's components. The car works in all normal conditions and even provides alerts in case of malfunction. Of course, it does not apply to the rock because it has no edge cases.

Does it have manufacturing protection?

Regarding manufacturing protection of the car, there are multiple stages of production and various tests that verify the quality of the car in the factory. Of course, one of the final tests is to check if the driver can drive the car. There are no verification tests for the production of rocks at all.

Does its product include protection from intruders?

Regarding protection from intruders, multiple systems in a car can ensure protection, including door locks, car keys, and immobilizers that protect the cars with a PIN code. Even a GPS tracker could be used to find a stolen car. However, in the rock, there is no protection from intruders.

Does it have recovery protocols?

Regarding recovery protocol, the car has multiple recovery procedures. There are procedures for engine maintenance, oil change, and gas refueling. Manufacturers also provide maintenance manuals and spare parts catalogs. To maintain the cars, manufacturers even produce components for an extended period of time. Some regulations require manufacturers to make spare parts for about ten years. Of course, there are no recovery protocols for the rock.

Does it have an adjustment of performance for the task?

Regarding performance adjustments to tasks, let's talk about cars. The car's acceleration depends on the throttle, which the acceleration pedal controls. The gear changes the engine's torque ratio to rotational speed depending on the car's speed. The torque is then transferred to the wheels. Multiple systems in the car require dynamic adjustments like throttle and gear ratio. Of course, rock does not adjust its performance.

Does it have an efficient design?

Regarding efficient design, cars are engineered to have low air resistance, which makes them more efficient. Engineers also ensure that the engine is as efficient as possible. Then, there is a battery, which saves some of the engine's energy. This battery allows the driver to start the engine easily next time. Engineers ensure the car's design is efficient and convenient for the driver. Unsurprisingly, because rock doesn't have a clear purpose, it's impossible to measure its efficiency.

Does it have the optimal location of systems?

Regarding the optimal location of systems, the car is a good example of well-placed components. Each component is thoughtfully placed. All elements in the car are placed in optimal locations: door handles, doors, seats, and the engine. For the rock, nothing is placed in an optimal location.

Does it have a complementary design?

Regarding the complementary design, there is no better example than the car. I call it a complementary design when there is a perfectly matching combination of symmetrical and nonsymmetrical elements for a clear purpose. Many components in the car are not symmetrical, which are related to oil, cooling, and electrical systems. On the other hand, the outside of the car is almost entirely symmetrical. This is because it reduces drag and improves the car's aerodynamics, making it more fuel-efficient. Even if the engine is not symmetrical, engineers would ensure the center of gravity is in the middle. Because the center of gravity is in the middle, the driver has predictable control over the car, which makes the car safer. The rocks would come in all shapes and forms. Some could be symmetrical, while others are not. However, rocks would not be complementary. Rocks would not be a perfect combination of matching symmetrical and nonsymmetrical designs for a clear purpose.

Engineering
versus Selection

I want to calculate the probability of humankind being engineered. Let's check if this method of calculation works by answering the second question out of three: "Is there a measurable difference between the engineering and selection processes?". The primary goal of this chapter is to find this difference.

Is there a measurable difference between the engineering and selection processes?

Does it have a clear purpose?

Regarding the clear purpose, let's see if there is a difference between selection and engineering processes. One of the most striking things in engineering is an easily observable clear purpose for nearly everything engineered. The way the selection process works is that some mutations would be beneficial, and they would expand throughout the entire population. Deadly mutations would disappear. Mutations that are neither harmful nor beneficial would stay within the population. Those mutations would occupy a proportional amount of the population even if the population grows. Because those mutations occur regularly within a population, almost every individual would have some combination of those neither beneficial nor harmful mutations. The selection process results would be a mixture of impactful and nonimpactful mutations. Having a measurable survival advantage in the selection process and having a clear purpose are different things. This method looks for things that have a clear purpose and have negligible effect on the survival of the individual. When I imagine the difference between the selection and engineering processes, the selection process will produce outcomes with many nonimpactful mutations. The engineering process result would produce outcomes with clear purposes, which often would not significantly affect survival. This means engineered mutations, which have no impact on negligible survivability, must have a way to be distributed to the entire population.

Does it have safety over-engineering?

Regarding safety over-engineering, one of the core purposes of engineering is to ensure the solution's safety. That means that instead of supporting the selection process of the convenient in-the-moment genes, engineers would do everything possible to ensure safety in the long term of the solution despite occasional mutations. This also means that engineering aims to acquire a very stable DNA code. During engineering, it is not required to select the best genes. At the same time, there might be some variation in the design to support changing requirements for survival. The main goal of engineering is to preserve the design's original purpose and long-term safety features. This means that engineering aims to have robust measures that ensure the solution's safety, including engineered behavior. Engineers must also ensure a robust and complete immune system that will provide sufficient survivability for the individual. From the selection process perspective, the optimal goal is to have as many mutations as possible to generate more mutations for the gene selection process. Many mutations increase survivability in the short term, negatively impacting survivability in the long term. The selection process can have alternative solutions to over-engineered safety. For example, one possible option is to increase the reproduction rate. In contrast, how does an engineer make trade-offs? Engineers would put maximum effort into individuals' safety and minimize the birth rate because of material use efficiency or possibly the value of life.

Does it have precise limits?

Regarding the precise limits, manufacturing relies on precise limits and margins. A car's engine has a very tiny margin of error. Some car parts have very strict precision requirements, while other parts do not require such precision. However, due to the manufacturing process, precision often exceeds requirements. Let's take a look at the selection process. There would be no impact on the survivability of the individual if some limits were not precise. This means some limits in the body could be bigger or smaller without a measurable effect on survivability.

Doesn't it have anything extra?

Regarding anything extra, during the engineering process, unnecessary parts in the final product would not be present. Unused parts would not exist because no engineer wants to waste material. In the selection process, there will always be cases of mutations. This means that if some extra part does not affect the individual's survivability, this part would exist in the population in relative proportion. Or even worse. If the population gets low and the unnecessary mutation is popular within this group, most of the group members would have this mutation, if not the entire population.

Doesn't it miss anything?

Regarding missing things. Engineered products must have operational components. This means the engineer would not miss any part, ensuring the product is fully functional. However, things are different regarding the selection process. Something would always be missing that would not impact the living organism's survival. This could be easily demonstrated in the car. Some bolts could be missing in the wheels, the window could be stuck, the headrest could be missing, and the car would still be operational. The car can also miss many other components without affecting driving ability. Similarly, many things could be missing in a living organism without a measurable effect on survival. The human body has a long list of parts that have a negligible effect on survival.

Does its design involve multiple systems?

Regarding design, which involves multiple systems. A very significant part of the engineering process is the ability to design complex interdependent systems. For example, an engine in the car can charge a battery using an alternator. After the engine is turned off, energy stored in the battery can start the engine again with a starter. From his seat, the driver can control all the car systems. The car must have a complex electrical and mechanical design to achieve this. The car is an engineering solution that has multiple systems with complex interconnection. From the selection process point of view, including multiple interconnected systems is a challenging task that involves numerous nonimpactful mutations that would not benefit the survival of the individual throughout a long chain of mutations. If the selection process created these complex systems, nonimpactful mutations would regularly occur within the population, even after the design is complete. Those nonimpactful mutations do not exist within the population. Therefore, there is a measurable difference between engineering and selection process outcomes.

Does its design include emergency options?

Regarding emergency options in design. If a car could reproduce like an animal, then during the selection process, the car would not be optimized for the safety of the individuals. One of the main purposes of the engineering process is to ensure that people are safe in case of an emergency. Engineered cars have all the safety features that operate specifically in an emergency. The seat belt, headrest, airbags, and crumple zone are all features engineered to ensure the safety of the individuals. They're not engineered to make sure that the car survives the crash. In the current example, the safety of individuals in the car represents all engineering considerations, while the car's survivability represents the selection process. From the perspective of the selection process, reproduction rate and survivability matter. However, this also means it's easier to increase reproduction than survivability. This is why safety would not always be chosen during the selection process. Over-engineered safety is undetectable from the perspective of the selection process. Over-engineered safety can only be selected when all individuals are continuously exposed to this specific type of danger. This means there is an expected significant difference between engineering and selection process results. It is important to note that engineered products would have robust and reliable safety features because they would be engineered and rigorously tested. This means that even though over-engineered safety features would rarely be useful, engineers would spend a disproportionate amount of time engineering and testing them. The selection process would have an issue selecting genes that would benefit survival in rare cases. Once the issue is rare, not enough individuals will be affected by the selection process that will filter safer mutations. This means the selection process outcomes with over-engineered safety do not have a chance to expand throughout the entire population.

Does its design work in all edge cases?

Regarding support of all edge cases. The main issue of the selection process is that it does not support all edge cases. Instead, the selection process chooses what works for most cases because it can only choose mutations with measurable survivability advantages. This means some edge cases with negligible effect on survivability would not be selected. During the engineering process, the core principle is ensuring that the product works in all edge cases. This means an expected measurable difference exists between engineering and selection processes in supported edge cases.

Does it have manufacturing protection?

Regarding manufacturing protection. Manufacturing facilities have multiple layers of manufacturing protection to ensure sufficient reliability. Engineers must ensure that the manufacturing process is reliable and production yield is high enough. However, expecting the same level of reliability in the selection as in the engineered process is unreasonable. This is because the engineering process would have very sophisticated measures that would prevent even hypothetical issues, not to mention rare issues with a low chance of occurring. The selection process can not foresee any hypothetical issues and can not select genes that are good enough to deal with rare issues.

Does its product include protection from intruders?

Regarding protection from intruders, engineers would analyze current and future threats and implement protection measures. This means that the product and its manufacturing facility would have sophisticated protection from threats, even in cases where those threats had never appeared before. Physical and digital access must have sophisticated security. From the selection process point of view, protection could appear only if threats consistently appear. This limits the selection process to only selecting genes that increase survivability against existing threats, not future ones. This is why complex, layered security measures are almost impossible to achieve during the selection process.

Does it have recovery protocols?

Regarding recovery protocols. Software engineering includes procedures to recover databases, files, and code. Recovery protocols also allow data restoration from snapshots in case of data loss. Engineers must always have backups for everything used in the system. During the selection process, recovery features would not make sense in many cases and would not be selected because those recovery features increase overhead on overall performance. By increasing the birth rate, the selection process can compensate for the lack of recovery features that affect survivability. Complex recovery protocols require a long chain of mutations that do not give any benefit for a long time. If the selection process is behind complex recovery features, then this active mutation process must result in vast quantities of nonimpactful mutations in the population.

Does it have an adjustment of performance for the task?

Regarding performance adjustments for the tasks, let's take my favorite car as an example. To control the car, the driver needs adjustable acceleration and deceleration. Selecting adjustable and complex systems is almost impossible during the selection process because they require a long chain of mutations. Those complex adjustable systems would need many nonimpactful mutations occurring regularly within the population. Only engineered solutions would consistently make multiple adjustable systems that work together for one goal.

Does it have an efficient design?

Regarding efficient design. There are multiple ways by which design may be efficient. There are several challenges in engineering, one of which is choosing the right energy source. Another challenge is engineering a very reliable solution that fits the requirements. Everything engineered must be efficient and reliable. The selection process rarely chooses the optimal energy source, most efficient solution, or most reliable solution because they are usually incredibly complex. Those complex solutions would not be efficient and reliable enough even after a dozen nonimpactful mutations, which makes them unreachable from a selection process perspective.

Does it have the optimal location of systems?

Regarding the optimal location of systems. Part of the engineering process is placing all the systems in wisely chosen locations. The location of a component affects the connection to other components, safety, and performance. During the selection process, due to random mutations, some body parts like teeth, bones, and ligaments would inevitably end up in random locations where they are not supposed to be. There would be a distinct pattern when some parts operate in weird places that are safe enough. However, those places would not be optimal because those parts were placed randomly.

Does it have a complementary design?

Regarding complementary design. Engineered products, when required, would have a complementary design, which means an efficient combination of symmetrical and nonsymmetrical design. Let's take, for example, my favorite car. Often, it doesn't make sense to create a symmetrical engine. So, due to the engineering process, the engine would not be symmetrical in most cases. However, considering the car's exterior, it would make sense to make it symmetrical mostly because of aerodynamics. This is because driving would be safer when the car's air resistance is symmetrical. The symmetrical exterior design also helps to reduce drag, which is caused by air resistance. For those reasons, some parts would be symmetrical while others won't. During the selection process, the anticipation is either a completely symmetrical or nonsymmetrical design but not a perfectly matching combination.

Origin of Humankind

B ased on previous findings, there is a significant difference in the engineering and selection processes. Because of this difference, it's possible to calculate the exact probability of humankind's engineering origin. Let's take parts of the human body and measure the probability using the method described in this book. There are many options because the human body has many incredibly complex parts, such as the heart, cardiovascular system, arms, joints, brain, teeth, bones, etc.

Taking any complex part and measuring the probability of engineering origin is possible. This calculation can be done on any complex part related to humans, animals, insects, and even machines like airplanes and cars. For instance, in the car, it's possible to take any part, such as the engine, battery, tires, and electric system, and measure the probability of the engineering origin. This doesn't mean that the selection process did not have any influence. Anticipation is that, like in the car, all parts are engineered. However, some parts were influenced by the choice of individuals, such as wheel covers.

The main goal of engineering is to create products that match the requirements. However, because engineering requirements change over time, ongoing projects require design iterations to meet changing requirements. The selection process is integrated into the evolution of cars and other products by changing requirements depending on what is popular. This is how engineers can choose the next improvements. I speculate that some body parts were affected by selection. However, the selection of engineered products differs greatly from the selection process described in evolution theory. I expect most components to have measurable engineering effort if humans were engineered. I want to clarify that I favor neither engineering nor evolution theories. I'm just curious whether humankind's origin is an evolution or an engineering effort. I'm driven by curiosity. Any option is good for me as long as I can pinpoint humankind's origin with nearly 100% accuracy.

Another thing I want to write about is that some people disregard Charles Darwin because they favor opposing theories. I see things differently. Can anyone imagine if Einstein would try to disregard Neuton? I see Charles Darwin as the forefather of modern biology. Nowadays, any new discovery stands on the shoulders of giants, including mine. The change that made Einstein was a revolutionary addition to Neuton's previous findings. I want to clarify that revolutionary findings were discovered only on the edge of the spectrum of Neuton's theory, which previously could not be verified. The same goes for me. I am adding a little step in a big journey, a little one on the edge of the spectrum. Charles Darwin's work is far more significant than mine. This is because Charles Darwin made a giant effort to collect information about species, which I greatly respect. And this was before the Internet when anyone could find anything. Charles Darwin collected data indicating that species transitioned from one state to another. This is how new species were formed. My effort can be considered as just a tiny rock on a huge mountain.

Let's answer the third question: What are humankind's odds of being produced by engineering effort? Let's take the heart and cardiovascular system for this calculation. Finally, I can answer this fascinating question. And this would be the final answer.

Does it have a clear purpose?

Regarding clear purpose. The heart and blood circulatory system have a clear purpose. They are responsible for delivering oxygen and nutrients to the cells in the entire body. The selection process must produce many non-impactful mutations without a clear purpose. Because the heart and cardiovascular system are complete, without parts with unclear purposes, the answer is "yes."

Does it have safety over-engineering?

Regarding safety over-engineering, the cardiovascular system has one crucial vulnerability. If a heart does not pump blood for a short time, the human brain begins to die. To maintain blood flow to the brain, the lungs must breathe, the heart must pump, and blood pressure must be within reasonable limits. Blood pressure requires very safe and robust control. Muscles inside the arteries expand and contract depending on emotional and physical stress. The nervous system controls muscles. The nervous system also redirects blood flow to organs that need it most and reduces blood flow to other systems depending on priorities. Inside arteries, muscles redirect blood flow to different organs by increasing and reducing the size of the arteries. For example, during exercise, arteries to the gut contract while arteries to muscles expand. The cardiovascular system is supposed to react within a reasonable time to maintain blood pressure. The nervous system ensures quick reaction time. Because control over contraction and expansion of arteries belongs to the nervous system, it ensures that the cardiovascular system will respond quickly to circumstances requiring more blood flow, like fights or dangerous situations. Once the danger is spotted, the nervous system increases its heartbeat and breathing. The nervous system controls the cardiovascular system and always ensures the body receives enough oxygen, even during dangerous situations. Heart and cardiovascular systems have safety over-engineering.

Does it have precise limits?

Regarding precise limits, like the car's engine, the heart has a precise design. To operate optimally, all of the heart's parts must have precise size relative to each other. The cardiovascular system is built in a way that all parts of the body receive oxygen. During the selection process, situations would occur when some body parts lack oxygen and do not operate at maximal performance. The heart and cardiovascular system have precise limits.

Doesn't it have anything extra?

Regarding anything extra. There is only one heart in the human body. There are some species with more than one heart, like giraffes. However, in humans, there is no unnecessary extra heart. I assume this is similar to why some aircraft have only one engine. In aircraft engineering, in some cases, it is neither economical nor practical to have two engines despite the safety benefits. In those cases, engineers would make additional effort to make the engine extremely safe and reliable. Engineers must ensure high reliability when designing something that impacts human safety. In some cases, engineers create a backup. In others, engineers make an extremely reliable product. The arteries and veins are all connected to the heart. There are no places where arteries or veins are disconnected from the cardiovascular system. Arteries and veins are not located inside ligaments. Ligaments are made from live tissue but do not require the bloodstream. This means that neither the heart nor the cardiovascular system has anything extra.

Does it miss anything?

Regarding anything missing. The cardiovascular system delivers oxygen and nutrition to the entire body without exception. Every finger, arm, and every other organ is connected to the bloodstream. Even bones are connected to the bloodstream, which is extraordinary. During the selection process, some body parts like bones and teeth would not be connected properly to the cardiovascular system. This is because not every bone or tooth is required for survival. The heart and cardiovascular system do not miss anything.

Does its design involve multiple systems?

Regarding interconnection with multiple systems. What is interesting about the heart and cardiovascular system is its collaboration with other systems. The nervous system collects information from organs, such as whether the body needs increased heartbeat or breathing. In the case of physical activity, the cardiovascular system redirects the bloodstream from areas that do not require lots of oxygen to the muscles that need oxygen the most. By doing so, the cardiovascular system can maintain pressure. During exercise, when muscles use more oxygen, the blood vessels inside of them increase in size, thus increasing the bloodstream. However, because the blood volume stays the same, other blood vessels decrease in size to maintain pressure. During exercise, blood vessels to the brain and gut shrink, increasing blood flow to muscles. The heart and cardiovascular system communicate with the nervous system. The conclusion is that the heart, cardiovascular, and nervous systems have an efficient interconnection.

Does its design include emergency options?

Regarding the inclusion of emergency options. The heart and cardiovascular system have safety features. For example, when an injury happens, it's vital to stop bleeding from the injury. This is done by sealing the wound with thrombocytes. To sustain life, the heart must always continue pumping the blood. This is why a heart will still operate even if the nervous system is disconnected. This is a reason why doctors can perform heart transplantation. People with heart transplants have a higher heart rate. The heart is so reliable that even without a nervous system connection, it elevates its heartbeat as a response to exercise. However, it does not respond as quickly as a heart with a healthy nervous system connection. People with heart transplants can still operate under normal conditions with some restrictions like contact sports. In an emergency, if the heart stops beating, there has to be a possibility to pump the heart manually. Because the rib cage is flexible, it's possible to pump blood through the heart by pressing on the rib cage. This also means the heart is conveniently designed to pump blood when pressed. The heart and cardiovascular system must operate without failures. If, even for a short time, there's no blood circulation to the brain, the brain begins to die. The bloodstream may stop for about two hours in some body parts like the leg or arm, almost always without damage. Humans usually sleep between 6 and 8 hours. Sometimes, a sleeping person can apply pressure that would prevent the bloodstream to an arm or a leg. In this case, a safety feature would wake the person because there is no bloodstream in one of the parts of the body. During the selection process, those safety features would not be possible to get. The heart and cardiovascular system design includes emergency options.

Does its design work in all edge cases?

Regarding support of all edge cases. The main engineering challenge with the cardiovascular system is the requirement to operate in a wide range of accelerations that affect the bloodstream. Those accelerations may happen due to gravity or by motion. Arteries under high pressure can operate in various conditions, but veins are not. Veins require a different design to operate in a wide range of accelerations because of negative pressure. Veins and arteries are connected through tiny blood vessels called capillaries. Blood under gravity might move in the opposite direction in veins and capillaries. The cardiovascular system is supposed to work from zero gravity up to earth's gravity with a large margin for physical activities. To make the cardiovascular system operational in all of those edge cases, veins and capillaries are protected by valves. Those valves allow blood to flow only in one direction. This allows the cardiovascular system to work reliably throughout various accelerations. Those valves allow the cardiovascular system to operate from zero gravity up to several times the earth's gravity.

Does it have manufacturing protection?

Regarding manufacturing protection. The main blood components are red blood cells, white blood cells, and thrombocytes. Without a sufficient amount of those cells, the human body would not be able to survive. But how is the manufacturing of those cells protected? The amount of engineering effort required to ensure the manufacturing protection of those cells is staggering. Where is the safest place to produce those essential cells? The bones are the safest place. The bones produce red blood cells, white blood cells, and thrombocytes. Many people may not know this, but bones are vital organs. The level of reliability of cell production is also insane. Red marrow inside bones is responsible for the production of those essential cells. However, not every bone contains red marrow. Bones also contain yellow marrow, which is fat that stores energy. The reliability of the production of those essential cells is based on the fact that yellow marrow and red marrow are interchangeable. If the production of blood cells is too high, then some red marrow converts into yellow marrow. However, when a person is injured and loses some of the bones containing red marrow, other bones detect low blood production and convert some yellow marrow back to red marrow to maintain sufficient blood production. Red marrow is protected against physical damage inside the bones and has higher protection against radiation. This is important because red marrow has a very active cell division process, which is very sensitive to radiation and physical damage. This means the red marrow is located in the safest place in the body. Manufacturing protection of blood is at a very high level. This allows the heart and cardiovascular system to operate reliably.

Does its product include protection from intruders?

Regarding protection from intruders. Through the blood, white cells protect the entire human body from intruders. Blood vessels are used as highways for the immune system. For example, if an intruder is detected somewhere in the muscle, a white cell can travel through the bloodstream and navigate to the intruder. It's amazing how white cells can squeeze through blood vessels. White cells transform into very thin discs that go through the walls of the blood vessels. What is also impressive is the ability of the white cells to travel against the bloodstream to attack the intruders. It's important to note that the immune system has over 50 types of cells protecting the human body against different threats like viruses, bacteria, fungi, etc. The immune system is very complex and robust. The heart and cardiovascular system are protected from intruders.

Does it have recovery protocols?

Regarding recovery protocols. The blood vessels have multiple recovery features. When a person has a cut in the blood vessel, bleeding must be stopped to prevent a heart attack. Thrombocytes can detect the cut and attach themselves to the walls of the blood vessel. After thrombocytes attach themselves to the walls of blood vessels, they change state to dendritic. The dendritic state is when a cell, instead of being smooth like a ball, becomes like an octopus, trying to catch other cells with tentacles. All cells that touch those tentacles are immediately glued. Thrombocytes that are glued also change state to dendritic and, by doing so, seal the wound. After the wound is sealed, recovery of walls of blood vessels begins.

Does it have an adjustment of performance for the task?

Regarding adjustment of performance for the task. Arteries and veins have a very complex layered structure. Layers in arteries are very different than layers in veins. There is a significant reason for the difference between arteries and veins. Arteries have only high positive pressure, and veins have mostly negative pressure. Pressure makes the arteries round. Those arteries have specific layers that are resistant to pressure. Different arteries have different layers depending on where they are in the body and their purpose. Arteries closer to the heart are more resistant to pressure, and the arteries further away from the heart have more muscle tissue that controls the size of the blood vessels. On the other hand, veins have a much bigger inner layer due to lower pressure, and they have valves that ensure that blood goes only in one direction. Depending on the blood vessel usage, different organs have different blood vessel designs, which depend on the purpose of those blood vessels. The muscles in arteries redirect blood flow to organs where oxygen is needed the most. This is an adjustment for the task.

Does it have an efficient design?

Regarding efficient design. The heart is optimized to operate in various conditions. It does not have great mechanical efficiency, but it is efficiently designed, meaning it can continuously and reliably operate for a very long time without stopping. It's important to note that scientists tried to develop artificial hearts several times. However, replicating the heart's reliability, which makes about 100,000 beats daily, is an extremely difficult engineering challenge. Over 70 years, the heart beats about 2.5 billion times. Control over the size of the blood vessels increases the efficiency of oxygen delivery to the organs. The blood flow during exercise drastically changes because of heartbeat elevation and very smart control over the thickness of blood vessels. Blood vessels to the gut and brain shrink during exercise, and blood vessels to muscles expand. This allows efficient redirection of blood flow to muscles. Overall, the design of the heart and cardiovascular system is efficient.

Does it have the optimal location of systems?

Regarding the optimal location of systems. The heart, lungs, and the main blood vessels are located in the safest place possible. The heart and lungs are inside the rib cage, which protects them. This location of the heart and lungs ensures that the sleeping person would not accidentally stop the heart by pressing on the chest. The rib cage is still light and flexible but firm enough to protect the lungs and heart during sleep. There couldn't be a better design for protecting the lungs and heart. Main arteries and veins are located in the middle of the body, their safest place. Overall, the heart and cardiovascular system are placed in optimal locations.

Does it have a complementary design?

Regarding complementary design. Complementary design combines symmetrical and nonsymmetrical design that is optimal for safety and performance. For example, the car's engine is usually nonsymmetrical, exactly as the human heart is nonsymmetrical. Blood vessels throughout the body, such as arms and legs, have the same blood flow capacity on the left and right sides. However, they are not symmetrical. Looking at arms, anyone would see that the outer shape of arms is symmetrical. However, veins are not symmetrical in the arms and the rest of the body. It is similar to a car when the outer shape is symmetrical. However, the internal wiring and components are not symmetrical. In the case of the cardiovascular system and heart, it's a complementary design that combines a symmetrical outer shape of the body with nonsymmetrical internal components.

Conclusion

After all of those questions are answered, the odds in favor of the engineering origin of the heart and the cardiovascular system are overwhelming. They are greater than 99.9%. Because there are 15 questions with "yes" and "no" options, the odds are 1 in 32768 that humankind is the product of evolution. The odds in favor of engineering origin are 32767 in 32768. This is an astonishing result in favor of the engineering origin of humankind. This is the final number for the heart and cardiovascular system, which is not affected by the size of the universe or by the number of parallel universes. I anticipate that the same results would be calculated for other human body parts. If the results are the same, then the odds in favor of engineering would exponentially increase, leaving no room for doubt. Even with such limited calculations, which were measured on the heart and cardiovascular system, it is crystal clear that the origin of humankind comes from the engineering effort. This is the final answer.

Repeating Calculation

I want to calculate the probability of humankind's engineering origin for several body parts to be confident in the results. For this purpose, I will use the most fascinating body part: the teeth. The teeth are extraordinary because of their complexity, blueprint perfection, and outstanding growth sequence.

I want to explain why teeth are so extraordinary. Teeth are very complex to create because of very tough requirements. Comparison with bones will explain why. Bones are incredible vital organs. The reason why bones are so incredible is because they are not empty. Bones are pretty much like skyscrapers that are full of people. Bones are solid structures built from calcium and phosphorus with billions of cells that live inside them. Bones are connected to all vital infrastructures. They have veins and arteries that go across the entire bone. Bones also have nervous and lymphatic systems. Each cell in the bone receives oxygen and nutrients and is also connected to the lymphatic system, which disposes the junk. Bones increase in size as the body grows. Bones also adapt to gravity and mechanical stress. Cells in the the bone change the structure and size of the bone depending on everyday stress and the amount of minerals. Bones are solid structures full of cells that repair, build, and disassemble the bone depending on requirements. Even nerves are found inside the bones.

The first reason teeth are so incredible is the conditions in which teeth must function. Teeth face scratches and repeated compression and decompression cycles, and their maintenance options are limited. The teeth structure is made from several components. The top layer is the hardest material in the human body, called enamel. Underneath the enamel, there is a layer called dentin. Both of those layers lack cells. Only the layer underneath those two has cells, which is called pulp. Teeth are more challenging to engineer because the enamel requires a unique maintenance method, which must provide lifelong durability in extreme conditions. Those conditions are beyond the properties of regular materials, such as zinc, gold, and ceramic. The solution must maintain teeth without cells inside the top layer.

The second reason teeth are so incredible is because the surface of the teeth is exposed. This is very different from bones, which are completely inside live tissue. This has big consequences because the exposed part of teeth has only a limited capacity for repair and cannot grow or increase in size. The teeth structure must match the blueprint as construction begins. After construction is complete, it's impossible to change the shape and size of the outside surface of the teeth. This means that, unlike bones, teeth are unique because the exposed surface of teeth must have a perfect shape at the beginning of growth. What makes teeth even more incredible is the fantastic sequence by which milky teeth grow and then are replaced by regular teeth. At first, milky teeth begin to grow from the front, and as the jawbone grows and makes more space, other milky teeth surface. Then, frontal permanent teeth continue their construction inside the jaw bone, and the roots of milky teeth begin to disassemble. Afterward, milky teeth are shed once no root is left. Then, permanent teeth surface. The same happens to all milky teeth sequentially from front to back. Milky teeth are small and designed for a child's small jaw. Permanent teeth are much bigger and have a precise size for the jaw of an adult. An interesting fact is that milky teeth and permanent teeth begin to develop during pregnancy. Milky teeth begin to develop during weeks

6 and 7, while permanent teeth begin to develop during week 14 of pregnancy. Another interesting fact: Right before a child sheds the teeth, two rows of teeth can be perfectly seen on the X-ray.

The third reason teeth are so incredible is their shape. Each tooth has a specific shape, matching its purpose depending on its location. This means the teeth must follow a particular blueprint, which changes depending on the location of each tooth. On average, humans have 52 teeth throughout life, which is ideal for calculating the probability of humankind's engineering origin. Children usually have 20 milky teeth during childhood, while adults usually have 32 permanent teeth.

Does it have a clear purpose?

Regarding clear purpose. Teeth have a clear purpose, and it varies depending on their location. Frontal teeth are used to cut, while rear teeth are used to chew food. Milky teeth are built similarly. Both milky and permanent teeth have a clear purpose.

Does it have safety over-engineering?

Regarding safety over-engineering. Teeth pose a danger to the tongue and cheeks during a bite. This means there has to be a mechanic that would protect the tongue and cheeks. For safety reasons, the tongue and cheeks must not intersect with the teeth during a bite. However, humans can accidentally move their lips, cheeks, or tongue during a bite. There is a reflex that protects the tongue and cheeks during a bite. Muscles pull cheeks and tongue backward to prevent injury. This means that there is an additional neurological protection layer that prevents damage to the tongue and cheeks. Using the nervous system and muscles to protect the mouth from teeth is safety over-engineering.

Does it have precise limits?

Regarding precise limits. A simple look at the teeth would show how perfect their shapes and sizes are. In engineering, there is a significant difference in the error cost depending on which stage mistakes occur: blueprint or execution. The blueprint has to be perfect because the error cost is exponentially higher than in execution. In the execution, there could be small mistakes that may not affect the performance and reliability of the product. For example, let's take the blueprint and the actual real house. When there is a missing wall or a pillar in the blueprint, it significantly affects the safety of the house. However, a two-inch pillar placement mistake would usually not affect the house's safety. Teeth have a perfect relative size compared to each other. Also, they have an ideal shape that allows them to function properly. This function of the teeth changes depending on their location. Some teeth cut the food while others chew. Teeth have small margins regarding size. If one tooth is slightly taller, it would cause an issue because it would prevent other teeth from functioning properly during chewing. Overall, teeth have precise limits.

Doesn't it have anything extra?

Regarding anything extra, in some cases during childhood, teeth can grow in the wrong place due to lifestyle, which would not allow them to be perfectly aligned. However, a good doctor can perfectly align them because of their perfect shape, designed to match one another. Even though the selection process cannot detect extra teeth in most cases, the number of teeth and their placement are perfect, as if they were engineered. Up until recently, people had wisdom teeth without any issues. However, during the last 3 million years, the brain enlarged in size. At the same time, the jaw decreased in size. Decreased jaw size caused a reduced space for wisdom teeth in the jaw for some individuals. I'll discuss this fascinating subject later on in the book.

Overall, people do not have any extra teeth, except for some cases of ongoing jaw bone size reduction.

Doesn't it miss anything?

Regarding anything missing. Even if some teeth are missing, a person's safety is unaffected. People can live daily life without missing one, two, or even several teeth. This means that the selection process is not able to detect missing teeth. However, if teeth were engineered, there would be a precise number of teeth regardless of how they affect survivability. This means teeth do not miss anything.

Does its design involve multiple systems?

Regarding the involvement of multiple systems. A significant component of teeth design is a maintenance procedure. The teeth's enamel is a hard material. However, it is vulnerable to acid. Acid causes demineralization of the teeth's enamel. The primary source of acid is bacteria. Bacteria use sugars and carbohydrates as food sources to survive and multiply. Some bacteria types produce acids by the end of the food consumption process. To prevent bacteria from spreading, the mouth is always washed with saliva. Saliva is a more alkaline substance, which is opposite to acid. Saliva has multiple components that kill bacteria. Saliva also breaks sugars and carbohydrates, the main sources of food for bacteria. Teeth maintenance involves multiple systems which must operate synchronously. The nervous system controls the release of saliva into the mouth. Smell, taste, or thoughts about food trigger the release of saliva into the mouth. Tooth design also includes a complex mineralization system that mineralizes teeth and repairs damage caused by acid. The nervous system that releases saliva in case of food consumption is a smart way to use several independent systems in one solution to

keep teeth healthy. Teeth maintenance involves multiple interconnected systems.

Does its design include emergency options?

Regarding support of emergency options. There are multiple emergency options in case of bacteria invasion. For example, an immune
response to bacterial infection. Another one is a cell replacement in
the pulp of teeth after the infection. In case of tooth damage, the
cells near the dentin layer can increase the thickness of the dentin
layer to restore the strength of the tooth structure. Cells in pulp also
eliminate dead cells and the remains of bacteria left after infection.
However, the most remarkable feature is the replacement of unique
cells at the microtube's entrance in the dentin layer. Those cells are
called odontoblasts, which can thicken the dentin layer if the structure of the teeth is damaged. How exactly cells that make repairs after
infection can find those tube openings and replace the killed during
infection odontoblast cells is a mystery. Overall, there are emergency
options for the teeth.

Does its design work in all edge cases?

Regarding support of edge cases. Teeth enamel is the hardest material
in the body. Enamel is so challenging from an engineering perspective
that scientists still cannot manufacture this fabulous material. The
engineering of the teeth involves providing a very durable solution.
To prevent scratching, teeth must be harder than the hardest origin
of food: the bone. The teeth are designed to withstand all typical
chewing and biting forces. This doesn't necessarily mean the teeth are
under no circumstances can be damaged as a result of physical damage or acid exposure. This means that the design of the teeth must
reliably function under normal circumstances. The teeth's design is

durable enough to last a lifetime with a proper mouth care routine and consumption of enough minerals.

Does it have manufacturing protection?

Regarding manufacturing protection. The teeth are never completed. Which means the teeth remain under construction throughout their entire life. While some parts of the teeth are maintained by cells, other teeth parts like enamel and dentin don't have cells inside to maintain them. Enamel is a super nanomaterial that has mind-blowing properties. There are several properties by which materials can be measured. The property that is very interesting for the teeth is hardness, and another one is fracture resistance. Hardness is the ability of a material to resist scratches. For example, glass for mobile phones is scratch-resistant. Hard materials usually have some bad properties that are extremely hard to overcome. They're very brittle. Once you drop your phone on the concrete, there's a significant chance that the screen will experience a catastrophic failure. It can shatter into pieces. Enamel has more hardness than copper or iron and has slightly less hardness than steel and titanium. It also has less hardness than ceramic. On the Mohs hardness scale, the enamel has a hardness of 5. For reference, copper is 3, iron is 4.5, titanium is 6, steel is 6.5, glass is 6.5, ceramic is 7, and diamond is 10. Enamel may not be the hardest material. However, two exciting properties make enamel much better than any other hard material. One of those properties is resistance to the propagation of cracks. Enamel has outstanding crack propagation resistance for hard material. The teeth have a much lower chance of experiencing catastrophic failure than other hard materials. On top of that, enamel has one exceptional property. Other materials used to replace teeth can last, in the best case, maybe 10 to 50 years. The enamel, with proper maintenance, is virtually indestructible. There is a problem with all hard materials that engineers can manufacture. They crack with cycles of compression and decompression. For example, the hull of 747 Boeing can handle 35,000 pressurization cycles. The same issue happens with any teeth implants. Materials for teeth available today are vulnerable to corrosion, microcracks, or both. Enamel is a super nanomaterial that can heal itself from microcracks and corrosion by the mineralization process. Beneath enamel, there's a layer called dentin. The teeth' unique dentin layer design allows them to pump through microtubes water with minerals. The structure of the enamel allows the repair of microcracks with those minerals. In the beginning, when teeth are formed, enamel construction starts when cells create a mostly hollow organic structure. Water with minerals goes in and out through the microtubes in the dentin layer. Those minerals are attached to the organic mesh and produce a crystallite enamel structure. Those minerals inside the enamel take up to 96 percent of the volume. One percent is left for organic structures and three for water. When exposed to acids, the enamel loses some portion of the crystallites. Because the construction of enamel is never over, water still flows through the enamel and continues to produce enamel crystallites. Enamel is not prone to microcracks. However, the mineralization process heals those microcracks. Those unique self-healing properties make enamel durable enough for lifelong use. I suspect enamel can last more than a billion years with proper maintenance. Proper maintenance involves removing acids by killing bacteria inside the mouth and consuming enough minerals to restore enamel. Teeth are protected from bacteria by saliva. This process of pumping minerals through enamel ensures continuous maintenance of the teeth. The permanent teeth are built inside the jaw bone. This protection ensures that the shape of the teeth will not be physically damaged. Teeth have manufacturing protection.

Does its product include protection from intruders?

Regarding protection from intruders. The teeth have multiple protection mechanisms against bacteria. The first layer of protection is saliva, which can destroy bacteria's walls and bacteria's food sources like sugars and carbohydrates. Saliva is also an alkaline substance that washes acids from the mouth produced by some bacteria types. Inside the tooth, there are also multiple protection mechanisms. One is in case bacteria destroys some portions of the enamel, and the second is if the enamel was damaged by physical force. Inside the dentin layer are located specific cells called odontoblasts that detect if the enamel is thinner than it should be. If enamel is too thin, those cells thicken the dentin layer to strengthen teeth and prevent exposure of those cells to bacteria. An immune response is triggered if bacteria penetrates the teeth and kills the cells inside. When it happens, the immune system kills bacteria. Then, cells are replaced, especially cells near the dentin layer, which are called the odontoblasts. They are located exactly at the opening of the microtubes. This means that multiple systems are involved in teeth protection from intruders.

Does it have recovery protocols?

Regarding recovery protocols. Teeth have several recovery protocols to function over a lifetime. One of the teeth's core properties is their maintainability. The most crucial component of the teeth is enamel. It is the material with the highest hardness in the body. Enamel has a higher hardness than nails made from iron or copper. Enamel's hardness is slightly below steel and titanium. However, unlike other hard materials, enamel looks like material from the future. Enamel is made from organic material that is eventually exposed to water with minerals. In the beginning, this organic material is mostly hollow. However, once exposed to water with minerals, everything changes. The minerals continuously attach themselves to organic structures and form crystallites. After that, nearly all of the enamel's structure becomes made of minerals. There is one vulnerability, and it is acid, which is produced by bacteria. This acid destroys minerals within the enamel. To cope with this problem, water with minerals is continuously pumped to the enamel through special microtubes in the dentin layer. This way, enamel is always maintained and recovered daily. This method also allows microcrack repair. Self-repair is a property that makes enamel the perfect material for the teeth. There is also another recovery protocol if the enamel is damaged. Cells called odontoblasts can measure the changes in enamel's thickness and thicken the dentin layer if needed. Teeth have multiple recovery protocols.

Does it have an adjustment of performance for the task?

Regarding adjustment of performance to task. Saliva release is adjusted depending on how much food is eaten. This saliva protects the teeth from acid produced by bacteria. Saliva is a core component in fighting bacteria and bacteria's food sources, such as sugars and carbohydrates. Numerous components in saliva act like antibiotics. The nervous system regulates saliva release by receiving information from smell, thought, and taste to treat incoming food. The next fascinating factor is the adjustment of the strength of the lower jaw. The lower jaw bone can detect if forces applied to it approach structural limits. When this happens, cells inside the jawbone reinforce the bone structure by increasing bone density. Multiple systems adjust their performance to maintain the teeth.

Does it have an efficient design?

Regarding efficient design. There are several layers in the teeth, each with a dedicated purpose. The teeth have hard, self-healing nano-material on the top surface, which is enamel. Underneath teeth also have a dentin layer equipped with microtubes, which supply water to the enamel to mineralize it. The teeth are not connected firmly to the bone. A shock absorption layer called gingiva or gum is located between the teeth and the bone. This design reduces microfractures in the teeth. The enamel's mineralization process efficiently repairs microcracks in the teeth. This outstanding ability allows not to replace teeth every five to ten years. Teeth are efficiently shaped. Frontal teeth cut the food, and rear teeth chew the food. The teeth are designed very efficiently.

Does it have the optimal location of systems?

Regarding the optimal location of systems. The teeth structure is very complex. Enamel is the hardest material in the human body. However, what is more important is that the enamel's thickest part is on the top side, which experiences maximum pressure. This means enamel has the highest thickness in the most critical spot. Enamel is thicker on top of the teeth and thinner on the sides of the teeth. Based on their function, different types of teeth are located precisely where they are supposed to be. This means teeth are located optimally.

Does it have a complementary design?

Regarding complementary design. In the previous example, the heart and the cardiovascular system were discussed. None of those systems were symmetrical, like many other organs. However, the outside shape of the human body is symmetrical. This is similar to engineered products, such as the internal components in the car, which are non-symmetrical, while the outside shape of the car is symmetrical. In the jaw bone, the outside shape of the jaw bone is symmetrical. However, the internal structure of the bone is unique for each side and changes over time depending on mechanical stress. It rebuilds itself over and over again while maintaining a symmetrical outside shape. The same applies to teeth, which are symmetrical on the left and right sides of the mouth. However, inside, they're unique. This means teeth have complementary designs.

Conclusion

Let's assume that only ten questions can be used to measure the difference between evolution and engineering. This will allow me to calculate about 1 in 1000 odds for each tooth, indicating that the selection process produced humankind. This is because all answers indicate that humankind is a product of engineering effort. For each tooth, the odds favoring engineering effort theory are 999 in 1000. Of course, 1000 is a rounded number from 1024. Because each tooth has a unique design and precise location, I can calculate the odds for 20 milky and 32 permanent teeth, which is 52 teeth in total. The result is 1 in 10 to the power of 156. Because there are three zeros for each tooth multiplied by 52, the result has 156 zeros. The odds in favor of the evolution theory are 1 in 10 to the power of 156. The odds in favor of engineering effort theory are 10 to the power of 156 minus 1 in 10 to the power of 156. This result is devastating for the evolution theory. This means the odds favoring the evolution theory are close to zero.

In nonscientific words, humans were engineered without any doubt. Simply put, there is zero chance that evolution created humankind. The odds in percentage in favor of the evolution theory are 10 to the power of minus 154. This number is very hard to imagine for the human mind. Let's imagine the odds as sizes of physical objects so it would be easy to understand. The difference in the magnitude of the size of the known universe and proton is about 10 to the power of 41. Let's try to imagine odds in favor of engineering effort theory as big as the size of the known universe. The odds favoring evolution theory would be smaller than proton many times. The eyes and even the best microscope would not see an extremely small object like that. This is not all. Let's imagine another universe as big as the known universe but scaled down to the size of a proton. The odds favoring the evolution theory would be smaller than a proton in that scaled-down universe. It's possible to do this even one more time. This is how small the odds are in favor of evolution theory. In simple words, humans were engineered beyond any reasonable doubt.

Proper Teeth Care

While watching lots of information on YouTube for this book, I found some interesting information about teeth care, which I think is useful to share. I was not able to confirm this information on myself. However, because it's potentially useful for readers, I'll mention it here. Disclaimer: I am not a doctor, so I can't give any medical advice. Please consult your healthcare provider before using any information described in this book.

Teeth care is necessary to meet most individuals' needs, and it involves brushing their teeth at least once a day, potentially even more. The second thing is crucial. Without minerals, it's not possible to mineralize teeth' enamel. Taking a complex of minerals is important for teeth care. However, this routine requires regular tartar removal caused by the calcification of biofilm produced by bacteria. Some people may think that food contains enough minerals. Usually, it is. However, it depends on the individual's nutrition and food quality. It's important to note that minerals are mostly lacking in food due to soil's continuous reuse on a production scale. Two additional options can improve teeth health if the usual routine is insufficient.

One option is white pulling. White pulling has been known for a very long time. However, it's important to note that it doesn't replace brushing the teeth. White pulling is just taking a spoon in the mouth of coconut oil and moving it inside the mouth for about 15 minutes after a meal. It could be another oil, but not all oils are effective. Afterward, it's important to spit it out because it's full of bacteria. White pulling uses specific oils to help kill certain bacteria types that build a biofilm barrier around themselves. White pulling destroys this biofilm barrier and helps saliva to kill bacteria. Saliva effectively kills bacteria by destroying its walls after those barriers are gone. It is important to note that most harmful bacteria types are efficiently killed by saliva without intervention. Only a handful number of bacteria types can resist saliva.

The second option is a special sugar called xylite or xylitol. It has about half of the calories of sugar. This substance is important for mouth care because good bacteria can process it, but bad bacteria like streptococcus mutans cannot. It doesn't work well for them. The best use of xylitol is candies, which you place in your mouth shortly after a meal. The spread of this sugar in the mouth helps good bacteria to replace bad ones. As far as I understand, this technique can almost eliminate bad bacteria in the mouth. However, it's important to note that consuming lots of xylitol is not good for health. It has half of the calories compared to sugar and is linked to clot formation in the blood because it affects thrombocytes. So, the best use of xylitol is candies after meals. It's better to avoid any other food products that contain xylitol. Please note that it can significantly reduce tartar formation because tartar is a calcified biofilm produced by bad bacteria types like streptococcus mutans.

Engineering Effort

I t is still a mystery where this engineering effort comes from. For this chapter, I'll refer to the origin of this engineering effort as computation. This means a series of decisions must be made for a good engineering result. However, who or what makes this computation? I'll leave it to your imagination. One option is that the human genome was changed at some point, and now computation has no influence on the genome. The second option is that this computation still affects the human genome. I don't know which is true, but based on my calculations, one of those options is true.

If a computation happened at a specific time, it is no longer observable. If so, the only evidence would be differences between engineering and evolution characteristics. However, if engineering effort still affects the genome, it's possible to detect it. It is also possible that scientists would find traces of engineering solutions that were used to change the genome. For example, evidence that could point out mechanics that allow computation to select specific genes. Surprisingly, scientists already discovered at least one mechanic that selects genes.

Evolution theory focuses on the survival of the fittest. In other words, the selection of genes that positively affect the survival of the individual. However, if computation is continuously involved in this process, it must be involved in gene selection. Or even more, what if this computation can produce gene changes? Even create new ones if needed. During the natural selection process, it simply cannot happen. So the question is: "Is it even possible to select specific genes at a higher rate than 50 percent?". Surprisingly, the answer is "yes". And there is a really good example of that.

In the world, the ratio between males and females is about 101 to 100. This means an almost exact 50/50 ratio of chromosomes XX to XY. Given this data, one would assume that there is a 50 percent chance of having an XX chromosome in a newborn baby. This is not correct. At birth, males and females have about a 106 to 100 ratio. This indicates that a computation affects the selection of genes in DNA.

How is this gene selection mechanic related to evolution theory? The "Survival of the Fittest" is based on the speculation that only genes that positively affect survivability would be present in the population because they eventually override all other inferior genes. This is based on the speculation that genes are passed to offspring at exactly a fifty percent ratio.

In the case of XX and XY chromosomes, gene selection deviates about 6 percent from the 50 percent assumed by the evolution theory. This is possible only if sophisticated computation consistently affects gene selection. This example indicates that gene selection is computed regularly.

Since this book proved that humankind was engineered, let's try to find out how this computation could affect the genome. Until this point, most people believed that evolution created humankind. Evolution theory is founded on the belief that genes are always passed at exactly a 50 percent ratio. Now, here is a section where I make speculations. Up until this point in the book, everything was solid and proven. From this point, I am asking a reader to be more forgiving. Please consider that the following is my speculation and not a solid fact.

What about the mystery of the extra wisdom tooth, which is no longer needed?

At first, I was baffled by wisdom teeth. Some people have them, and some people don't. The only issue with wisdom teeth is that some people have a condition called impacted wisdom teeth. It doesn't happen to everyone but to people with a smaller jaw. This condition is when the jaw is so small that wisdom teeth grow in the section of the jaw bone that bends upwards. This way, wisdom teeth grow at an angle and push teeth forward. Consequently, the top of the wisdom tooth pushes a side of the teeth near it. In some cases, it could lead to pain, inflammation, and even potentially death. Not every wisdom tooth is life-threatening. When the jaw bone is large enough, wisdom teeth grow as they should without any issues. However, some people don't have wisdom teeth, and some miss one or more wisdom teeth.

I wondered what caused the issue and how computation could be involved if it still affects the human genome. To explain my thoughts, I need to present information about wisdom teeth.

What causes impacted wisdom teeth?

- The continuous decrease in the size of the jaw bone.

What can cause a decrease in jaw bone size?

- The continuous increase in brain size over the last 3.5 million years.
- Breathing through the mouth drastically affects the jaw bone.
- Soft food causes a decrease in jaw bone size.
- Natural selection in case a smaller jaw bone has survival advantages.

What are the consequences of a decrease in the size of the jaw bone?

- People who have wisdom teeth may have health issues
- Some people may die as a consequence of health complications.
- It may lead to the selection of genes that suppresses wisdom teeth formation.
- In rich countries, impacted wisdom teeth will have a negligible effect on survival.

What can cause wisdom disappearance?

- A combination of many genes affects the formation of wisdom teeth.
- Anesthesia in gum between ages 2 and 6 suppresses wisdom teeth formation.

Who doesn't have one or more wisdom teeth?

- Indian people: 11.5%
- Korean people: 41%
- Indigenous Mexicans: 100%
- The worldwide population: 22%

When was the first recorded case of impacted wisdom?

- About 13,000 to 15,000 years ago

When did people start to produce food by agriculture?

- Agriculture began about 12,000 years ago

When was the first recorded case of missing wisdom teeth?

- Between 200,000 and 300,000 years ago

I took this data from the Internet without performing a deep fact-checking. There could be some mistakes in the data. However, it will be enough to explain that multiple factors play a role in the presence or absence of wisdom teeth, and genetics plays a significant role in them.

I've got speculation about how computation is involved. To explain myself better, I want to use an analogy of how I write this book: I write a chapter and resolve all Grammarly alerts. Then, I upload the chapter to Drive as a PDF. While I walk somewhere, I review the chapter and add comments. There could be about 200 comments in a big chapter for the first time. Then, I return and insert all the changes into the text and resolve all the Grammarly alerts. The new version replaces the old one in Drive. Then, I'll repeat the process. At first, there would be about 200 comments, then 150, 100, 50, 25, 15, 10, 5, 2, 1. The number of comments will decrease each time I read and fix the chapter. However, not everything is so simple. Sometimes I will have an issue. The information can be incorrect, or I find a mistake. There could even be a contradiction between paragraphs in the chapter. Then, I'll rewrite the section with a mistake. The number of comments representing corrections will increase for this chapter section. The comments will gradually decrease once I resolve the issues: 15, 10, 5, 2, 1. I can change a few words, and as I read the chapter again, I can accidentally find some inaccuracies that affect other paragraphs and even other chapters. I will do that until I have zero comments after reading. I know it is inefficient, but it's just how I do it.

How is the way I write the book connected to the human genome? Well, computation rewrites the genome in the same way. It reads the genome, finds the problematic part, and rewrites the section to solve the issue. Scientists perceive this rewritten section of the genome as a variation in a gene. After that, the computation examines the solution to see if the problem is resolved. This examination probably happens throughout the life of individuals. After that, the gene refinement process gradually happens. Then, the number of variations decreases as the solution becomes more refined. Even though it is highly speculative, this is how I think computation affects the genome.

If my speculation is correct, it could explain the variations that the great Charles Darwin first saw. It also explains why some DNA parts are stable while others change constantly. If it's correct, then it's possible to predict which parts of DNA will experience significant variation.

It's possible that several options can trigger DNA variation. One is genetic, in case changes in the gene cause issues in other parts of the DNA. It is also possible that the environment causes engineering issues, and then computation tries to resolve them. It could be the first, second, or both options. If there is an environmental factor in the variation of genes, then it would be possible to use it for agriculture. I speculate that the challenges crops face are used in computation to produce variations of related genes. It also causes changes in parts of DNA that must be corrected because of changes in those genes. I speculate that physical or chemical interference could be used to cause gene variation, allowing the selection of more productive crops.

Regarding the mystery of wisdom teeth, not every person requires the suppression of wisdom teeth growth. Many factors may affect wisdom teeth, including consumed food and breathing. Assuming my speculation is correct, if wisdom teeth cause an issue, an increase in variation will occur that affects genes associated with this issue in an attempt to resolve it. I speculate that this is why some people miss one or more wisdom teeth. Then, if the issue is resolved, the number of variations will decrease.

Conclusion

After this book proved that engineering effort exists in the human body, the only question left is where this engineering effort comes from. It is still unclear whether this computation affects the genome nowadays. However, I believe scientists will find the origin of this engineering effort in the future. I think my speculation about the computation effect on the genome is the biggest blunder in this book, but it is the best I can do.

I have made so much effort to make it possible for you to read this book. I hope this book is helpful to you. This book was not designed to be entertaining. It was designed to be accurate and articulative. I want this book to make a difference in the hearts of billions of people worldwide. I believe they need it. If you can help spread the word, any help would be appreciated.

The End

———

Thank you for reading this book. I hope it serves you well. All I want is to make a change for billions of people who want to know where we came from. I believe everyone has the right to know the truth. If this book does a great service to you, I will be very grateful to serve you. Thank you.